IMAGES
of America

SYRACUSE

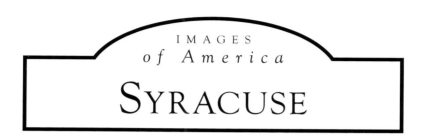

IMAGES
of America

SYRACUSE

Elizabeth A. Najim

ARCADIA
PUBLISHING

Published by Arcadia Publishing
Charleston, South Carolina

Printed in the United States of America

Library of Congress Control Number: 2023941787

For all general information, please contact Arcadia Publishing:
Telephone 843-853-2070
Fax 843-853-0044
E-mail sales@arcadiapublishing.com

Visit us on the Internet at www.arcadiapublishing.com

To my father, Harry L. Najim. I wish you were here to see this book come to fruition. May your memory be eternal. To my mother, Sue E. Najim. Thank you for introducing to me a love of books and museums. All my love, always.

CONTENTS

ACKNOWLEDGMENTS

As a history- and photography-oriented book, Images of America: *Syracuse* would not have been possible without the resources at the Syracuse Regional Museum, the Utah State Historical Society and its online digital library, the archives at Weber State University, the resources at Antelope Island State Park (thank you Carl Aldrich), guidance from tribal elder Darren Parry, and various museum volunteers (especially Diane Palmer) who assisted with their own contributions, both historical and photographic. A thank-you is owed to my editor Amy Jarvis for assisting me throughout this process. A special thank-you goes to my Najim family and friends who kept me motivated during the tougher months of this project, as well as my German shepherd Leo for his endless love. Unless otherwise stated, all photographs are courtesy of the Syracuse Regional Museum.

INTRODUCTION

The land to the east of the Great Salt Lake has seen a variety of peoples, cultures, and lifestyles. Native American tribes, specifically the Northwestern Band of the Shoshone Nation, originally settled the land. Mountain men came as early as the 1820s, including notable figures such as James Bridger, Jedediah Smith, Kit Carson, and John C. Fremont. They searched for streams that flowed into the lake in hopes of finding beavers to trap and other wildlife to hunt.

The Old Emigrant Road is one of the earliest historic spots in Syracuse. The road started in Salt Lake City and ended in the City of Rocks, Idaho. This road, meant to be a shortcut through Utah to those seeking gold in California, served many different groups for over 140 years. It benefited emigrants looking for a new place to call home with better foraging and water for their animals, avoided the sand hills to the east, and provided a smoother road surface for easier wagon trips. Today, it is known as "Bluff Road" in Syracuse.

With the Homestead Act of 1862, land in Utah became available for settlement. Any citizen could obtain government land in 80-to-160-acre claims. While many pioneers came and settled the neighboring areas, David Cook first plowed the land here in the spring of 1887 and sowed grain that fall when he was 20 years old.

In 1935, a group of residents met to consider appointing a town board. With the Great Depression and local issues such as unsanitary sewage facilities and unsafe wells, citizens wanted the benefits that an incorporated town would bring. The Syracuse Town Board organized in 1935, and the first elected members took office on September 11, 1935. They were Wallace Christensen, Thomas J. Thurgood (president), Lionel E. Williams, Elton Bennett, and T. Joseph Steed.

World War II brought many changes to Syracuse. While the United States never experienced battles on its home territory (except the attack on Pearl Harbor on December 7, 1941, which officially brought the United States into the war), the effort to support the military was at the forefront of every Utahn's mind. Syracuse was no different—those left behind tended to the farms, the canning factories, and their victory gardens. Many local men shipped off to the European or Pacific theaters.

The farming industry faded away as jobs at the new Hill Field air base and Freeport Center promised good pay and benefits. The war paved the way for new industries to build and expand in Utah, and the agrarian community slowly transitioned to a budding Salt Lake City suburb.

Syracuse's population remained steady until the latter part of the 20th century. The community was tight-knit; everyone knew everyone, and local sporting events, holidays, and religious activities brought people together. In 1980, the population was 3,702; in 1990, it was 4,658. By 2000, it was 9,398, and by 2010, it had jumped by 150 percent, to 24,331. Syracuse continues to grow to this day.

The transition to the 21st century created many changes for Syracuse. As more people moved to Utah, many settled in Syracuse, as it has always been a safe and affordable place to live. In March 2020, the COVID-19 pandemic struck the town as it did all over the world. People quarantined for weeks, schools and businesses shut down or moved to remote work, and face masks became required. Many from larger cities out of state moved to more rural places in Utah for a change of scenery. Syracuse was one of these places that experienced enormous growth in the early 2020s.

In its early history, Syracuse knew no boundaries, only landmarks. Today, Syracuse is rapidly changing from a farming community to an urban town. Old-time residents have mixed feelings about what this might bring, but as the saying goes—"You can't stop progress, only give it good direction."

One

Early Peoples
and Settlers

The Homestead Act of 1862 allowed any US citizen the right to obtain government land, usually in 80-to-160-acre claims. In 1864, the act was amended to grant to the railroads 10 alternate sections per mile within 20 miles on each side of the railroad line. Tracts of land in Davis County fell within these limits and became the property of the Union Pacific Railroad in 1869. This act, which divided up what is now Syracuse into different parcels, brought many new settlers to the area.

As the Hooper Canal reached Syracuse, eventually the usage of the land changed from grazing animals to dry farming, and eventually irrigated crops. It took immense effort to clear the land of greasewood, thistles, cactus, rabbitbrush, and sagebrush. The first pioneer to penetrate the Syracuse wilderness was a young man named David Cook. He built a log cabin in 1877, and his family eventually joined him from Salt Lake City in 1880. Migration of other families to the area continued throughout the 1870s; however, for many years, the area remained lightly populated, primarily because of a lack of potable water.

William Gailbraith and his wife, Phebe, came to the area in 1879 to develop a small homestead, although his main focus was developing a salt industry on the shores of the Great Salt Lake. He bought out George Payne's salt interests along with about 105 acres along the shoreline south of his homestead. For the next few years, he produced some of the finest salt west of the Mississippi.

Galbraith adopted the brand name Syracuse, named after Syracuse, New York, where the purest salt in the world was being produced at the time. In as little as two years, the name had stuck for the new city, too.

As pioneers settled and tamed the land, more people followed in their footsteps. These homesteaders, in search of a new life full of opportunities, happily settled in this beautiful part of Utah and eventually left a legacy of what Syracuse would become.

SHOSHONE INDIAN VILLAGE.

This print illustrates a typical Shoshone encampment in the 19th century. The Shoshone in this area were nomadic and frequently relocated between northern Utah and Idaho depending on the seasons. They are considered the first people to inhabit this land. (Used with permission, Utah State Historical Society.)

Chief Washakie (center) spent a lot of time in the Salt Lake City valley and was a friend of Brigham Young. He was the chief of the Eastern Band of Shoshone in the Wind River Mountain Range in Wyoming. He was the only Native American chief to have an Army fort named after him, Fort Washakie, which shows how much respect the Army had for him. (Used with permission, Utah State Historical Society.)

This photograph of Shoshone shows different types of leadership through dress. The Shoshone are known for their Plains horse culture. They acquired the horse in 1700, which greatly improved their lifestyle. It aided their nomadic nature and helped them to become proficient hunters and fierce warriors. (Used with permission, Utah State Historical Society.)

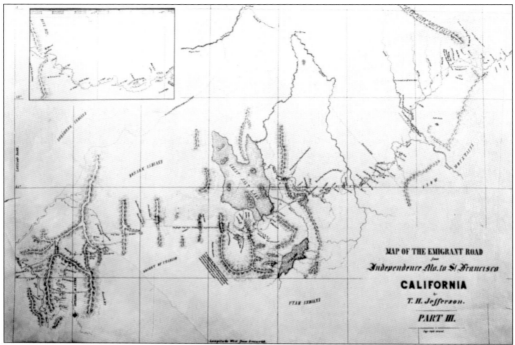

MAP OF THE EMIGRANT ROAD
from
Independence Mo. to St. Francisco

CALIFORNIA
by
T. H. Jefferson.

PART III.

The Old Emigrant Road was also known as Bluff Road, Old Traveled Road, and the Salt Lake Cutoff. It was the most practical way to reach the California Trail from Utah. Hasting's Cutoff, the alternate route, had proved impractical for wagons and livestock. Local residents used this route for many years, as it was the easiest way to travel in the valley. The road was first used by Capt. Samuel Hensley with 10 men in August 1848. On his advice, Mormon Battalion members returning from California also traveled this road. In 1849–1850, an estimated 22,500 gold seekers followed this route to the California goldfields. From 1852 to 1857, emigrant families used the road on their way to Oregon and California. The Rachel Layton Warren Camp of the Daughters of Utah Pioneers erected the historical marker below at the site of the trail in Syracuse in 1989. (Both, used with permission, Utah State Historical Society.)

The Mormon Trail was the 1,300-mile route from Illinois to Utah that early members of the Church of Jesus Christ of Latter-day Saints (LDS) traveled from 1846 to 1868. About 70,000 Mormon pioneers made the trek on foot, in wagon trains, or handcart companies to "Zion," hoping to find a home where they could practice their religion without persecution. Many of Syracuse's original settlers traveled to Utah this way. This photograph shows part of the trail in Wyoming. (Used with permission, Utah State Historical Society.)

While the term "spinster" usually refers to an older, unmarried woman, or a woman who is considered unlikely to ever marry, it originated in the Middle Ages to describe women who spun yarn. Many pioneer women, especially older ones with grown children, took up spinning thread and yarn to help make clothes or other cloth items. It was an ideal job for older people, as they could sit while working. (Used with permission, Utah State Historical Society.)

East of the Bluff		
Date	Settler	Address
1878	Richard Hamblin	1100 West 700 South
1878	Thomas J Thurgood	2278 South 1000 West
1878	Richard Venable	1699 West 700 South
1881	Peter Christensen	2098 West 700 South
1882	Alma Stoker	1452 West 1700 South
1883	James Warren	2550 West 700 South
1884	Amasa Driggs	1400 West 700 South
1890	Thomas Briggs	1275 South Bluff Road
1891	David Gailey	740 South 2000 West
1892	Ebenezer Williams	1602 West 2700 South
1893	Cyril Call	288 North 2000 West
1895	Charles Barber	1924 West 2700 South
1896	James H Baird	2455 South Bluff Road
1897	George H Bennett	1031 West 2700 South
1899	Joseph Willey	2955 South Bluff Road

The first settlers east of the bluff, sometimes referred to as Sand Ridge or Starvation Flats, came in 1878. Though similar to the surrounding area, settlers faced sandier soil and hotter conditions than those found in West Syracuse. It was ideal land for grazing. By 1900, sixteen families occupied the area. When they first arrived, they had to haul water in barrels on wagons to provide for their needs. Most pioneers settled west of the bluff initially due to better water sources until the Hooper and Layton canal systems were constructed.

West of the Bluff		
Date	Settler	Address
1877	Joseph Bodily	1740 South 4000 West
1879	Joseph Hadfield	4341 West 1700 South
1879	David Kerr	4258 West 1700 South
1879	George Payne	1200 South 4500 West
1880	William Beazer	3178 South 3000 West
1880	James Walker	3552 West 1700 South
1883	Ephraim Walker	3200 South 3000 West
1885	John Criddle	1140 South 1000 West
1885	George Rampton	1558 South 2000 West
1885	Henry Rampton	2750 West Gentile Street
1887	John Coles	3530 West 1700 South
1887	William Criddle	924 South 4000 West
1887	John Singleton	4700 West 700 South
1888	Daniel Walker	1797 West 1700 South
1888	David & Fannie Cook	1444 South 4000 West
1890	David & Hannah Cook	2600 West 1700 South
1890	William Miller	2640 South 4000 West
1890	William Bentley	958 South 4500 West
1893	John King	2025 West 2700 South
1893	Erastus Fisher	2177 West 1700 South
1897	Edward Tree	915 South 4000 West
1898	Homer Walker	1876 West 1700 South
1899	Fredrick Tree	923 South 4000 West
1899	Joseph Alexander	1375 South 4000 West
1900	Thomas E Williams	2517 South 2000 West

Patty Bartlett Sessions, one of the original Mormon pioneers who came across the plains in search of Zion, was a prominent midwife in the local area. Her diaries provide historians with an idea of life as a pioneer in the 1800s. She recorded 3,977 births in her writings, and averaged approximately $2 per birth. She also studied horticulture and set up a school in Bountiful. (Used with permission, Utah State Historical Society.)

This copy of *The Family Physician* (1842) belonged to Patty B. Sessions. Locals considered her to be the unofficial doctor of the Davis County area. She frequently consulted this text during her career. The book is on display at the Syracuse Regional Museum.

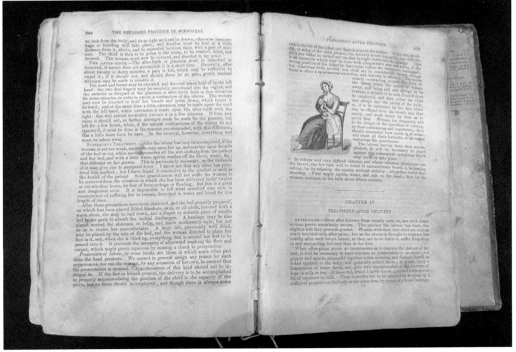

George Henry Payne was one of the first settlers in Syracuse. He became the first to open a salt business along the Great Salt Lake in Syracuse. In those early days of harvesting, salt deposits were made by evaporating the salt water. Payne did this from the ponds on the south side of the Syracuse road. Eventually, he sold his business to William Galbraith, who is known for effectively naming Syracuse through his salt business.

William Galbraith bought land along the shore of the Great Salt Lake and developed salt ponds into a thriving salt industry, producing as much as 20,000 tons of salt each year. Galbraith adopted the name Syracuse Salt Works for his business.

Known for the first furrowing of Syracuse land in 1876, David Cook served his community and was an active member in the LDS church. He moved back to Syracuse in 1890 after serving a mission in England, and became bishop of the South Hooper Ward on June 4, 1893. He was noted as a capable and well-loved leader.

James Thomas and Christine Cook Walker pose with their daughters Mary Golda (left) and Martha Christine. James Walker and his brothers Daniel and Ephraim started the Walker Brothers store at 3666 West Antelope Drive and operated there until about 1920. The first telephone line ran from Hooper to that store, which was the only telephone in Syracuse. Golda and Martha took the calls and delivered messages on horseback. Early in the 1900s, the brothers started a second store on the corner of 2000 West and Antelope Drive. Daniel managed it until about 1931, when it was bought by T.J. Thurgood, Samual Cook, and others. James was also a founder of the Syracuse Canning Company on 4000 West near the railroad spur that ran to the bathing resort. Christine was the sister of David Cook, in the previous image.

The prominent William Criddle family pioneered the Syracuse area in the late 1800s. They were among the first to settle west of the bluff. Their legacy continues today, as the family expanded. From left to right are (first row) Myrtle and Dora; (second row) Elda Viola, Sarah, Carl, William, and James; (third row) Lawrence, Millie, Delbert, and Della.

The John Rentmeister family, who originally emigrated from Europe and settled in Syracuse in 1901, was another well-known family in Syracuse's early history. Many of the children married Criddle family members. From left to right are (seated) Josephie, Agatha P., Paul, John, and Joseph; (standing) Mitchell, John B., Eli E., Rachel, Peter J., Antone J., and Frank.

This log cabin, known as the "Wilcox Cabin," was built in the 1850s in the Kaysville area. It was relocated to Syracuse sometime between 1879 and 1883. William and Emily Wilcox (below) moved into this little cabin on March 4, 1905. William purchased the home along with 160 acres from Christopher Layton on May 20, 1885, for $1,800. They moved in with a few of their belongings as they slowly built a new home for their family. They had eight boys and two girls. The family moved out of the cabin on Christmas 1911 as their new, larger home was ready.

The children of William and Emily Wilcox are, from left to right, (seated) Darvil Wilcox, Louise Wilcox Johnson, and Myron Wilcox; (standing) Lynn Wilcox, Elmer Wilcox, Mary Wilcox Miller, David Wilcox, Harold Wilcox, and Hugh Wilcox. Spearheaded by son Elmer, the family restored the historic cabin of their parents. Harold and Elmer were born in the cabin, on September 5, 1908, and May 5, 1910, respectively. The cabin currently resides at the Syracuse Regional Museum, a tribute to pioneering spirit.

Syracuse Utah Population Estimates 1890-2020	
Year	Population
1890	299
1900	299
1910	553
1920	629
1930	890
1940	732
1950	837
1960	1,061
1970	1,843
1980	3,702
1990	4,658
2000	9,398
2010	24,331
2020	32,141

Syracuse maintained a steady population with minimal growth from the first official census in 1890 until the post–World War II era. Throughout the 1980s to the present day, the population experienced extreme growth due to an increase in housing construction. Most of the Salt Lake valley experienced similar growth.

Two

A Farm Community
Is Born

Despite the high-altitude terrain with bitter-cold winters and hot summers, land in this area provided a hospitable environment for farming. The fertile soil near the Great Salt Lake gave ample opportunity for raising different crops. Hay and grain were the staple crops first grown for livestock, as feeding farm animals was necessary. Gradually, pioneers introduced other crops with relative success, including sugar beets, tomatoes, onions, potatoes, peas, and peppers. By 1900, Syracuse was the largest produce supplier in Davis County.

While early settlers took advantage of existing springs and streams, digging ditches and building small dams to water their land, the dryer seasons made this difficult. A more reliable water source was necessary for large-scale farming and homesteading. The Hooper Canal, begun in May 1872, moved water south to western Weber County from the Weber River. This, as well as the Davis-Weber Canal, was crucial to Syracuse's growth as a farming community.

The sugar beet ruled the land in Syracuse. The fertile soil allowed this crop to thrive, making it Syracuse's number one cash crop from the late 1800s until just after World War II. As farming technology improved, so did the harvest, as people no longer had to dig, top, and load the beets by hand. Even the cows could eat the leftover beets that were not deemed worthy of being sold. Sugar beets were shipped to the Layton Sugar Company to be processed into sugar. Each beet yielded about one cup of sugar. Many farmers invested in this industry and helped each other when times were tough.

Syracuse continued to be a thriving farm community well into the 20th century; however, the entry of the United States into World War II drastically changed the landscape for Syracuse, and farming became less of a thriving concern after the war.

Construction for irrigation systems began here at the mouth of the Weber River in 1881. After workers dug ditches by hand, they placed wooden planks to divert water from the river. Despite the construction occurring in warmer months, the Weber still had cold water and strong currents, making the project difficult. (Special Collections Department, Stewart Library, Weber State University.)

Construction of this canal, especially the Hooper Canal segue leading to Syracuse, was long and arduous. Workers relied on primitive tools such as shovels, picks, horse-drawn scrapers, and hand plows. Local farmers also worked on the canal, earning shares of stock for their labor. (Special Collections Department, Stewart Library, Weber State University.)

By 1875, the canal irrigated 5,000 acres of Syracuse. The Davis-Weber Canal, built in the 1880s, also irrigated the community. Residents set up ditch companies to ensure proper allocation of the water. With canals, farmers and citizens could properly farm and garden in this arid climate. The western United States has always had controversies regarding water control and drought restrictions, yet there is something to be said about those who constructed these canal systems to allow for growth. (Both, Special Collections Department, Stewart Library, Weber State University.)

A farmer, looking west, tends to his sugar beet farm in summertime (note the absence of snow on the mountains). Farmers harvested their sugar beets in the fall. (Special Collections Department, Stewart Library, Weber State University.)

Two young boys, looking quite underwhelmed, pose with a giant pile of sugar beets in the loader behind them. Most children, when not attending school, helped their parents with the harvest. Many continued to work on family farms as adults, eventually inheriting them.

Beet farmers worked long hours during harvest season. It was a laborious process that involved back-bending work. Farmers used beet topper tools to remove the foliage off the beet and dig the beet out. As technology improved, specialized machines helped with this process. (Special Collections Department, Stewart Library, Weber State University.)

Harvested sugar beets went to local beet dumps, near shipping areas, to be taken to the sugar factories for processing. A round trip from the farm to the beet dump took about an hour and a half depending on how far away the dump was and how long the line of waiting wagons was at the beet dump.

Due to the rich, fertile soil near the Great Salt Lake, sugar beets became the top cash crop in Syracuse. The majority of beets were taken to the local sugar processing plant. Generations of pioneers and their descendants made their living this way.

Farmers in Syracuse and Davis County used trucks to haul sugar beets to the beet dump. This 1929 Ford Model AA, restored by Clarke Fowers, served the Syracuse area for much of the 1930s–1950s. It is currently on display at the Syracuse Regional Museum. (Used with permission from *Wall Street Journal* photographer Chad Kirkland.)

By the 1900s, sugar beets kept Syracuse's economy thriving. Farmers had to get creative with harvesting these giant vegetables, which grew as big as 25 pounds each. The majority of beets were hauled to the local sugar processing plant. After harvesting, beets were placed for temporary storage in large piles by machines known as "beet pilers," consisting of a receiving hopper, a belt conveyor, a screen to remove soil and other debris, and a large boom that deposited the beets in piles.

The Layton Sugar Company opened in 1915 and employed up to 500 men, making it one of the largest factories in northern Utah. The factory and warehouse were built to house up to 110,000 bags of sugar. Farmers all over Syracuse shipped their sugar beets, mostly by railcars, to be processed into sugar. This facility was crucial to the success of the sugar beet crop, which was Syracuse's highest-yielding export. The factory closed in 1959 after World War II caused a slow end to the sugar beet industry. (Used with permission, Utah State Historical Society.)

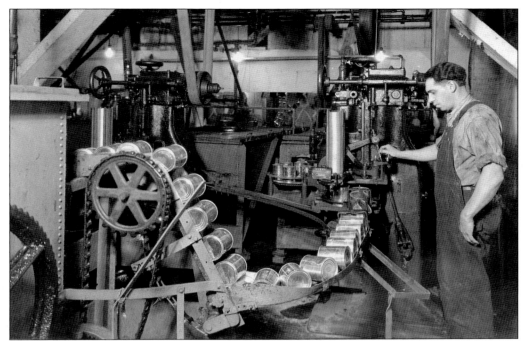

Canning is a method of food preservation in which food is processed and sealed in airtight containers, such as glass jars. While this method began in France in the early 1800s, pioneer families in Syracuse adopted canning as early as the 1860s to help preserve the surplus food from their crops. Canning helped to preserve food one to five years past its usual shelf life. Eventually, factories were developed to expedite this process. (Used with permission, Utah State Historical Society.)

Syracuse farmers grew potatoes as early as 1894. The fertile soil suited this crop. Most farmers grew potatoes in their own gardens and had their own cellar for winter storage. Eventually, potatoes became a staple crop in Syracuse and were shipped east on railcars for export. The average price was $1 per 100 or $20 per ton. Despite their excellent growth in this area, potatoes never gained the status of sugar beets or onions. The farm truck above is the 1929 Ford Model AA seen on page 28. (Both, used with permission, Utah State Historical Society.)

The Miller farm in Syracuse was known for its apple orchards. Many men were hired to pick during the apple season, and as business grew, so did the space for apple storage sheds. Pictured at the Miller Farm by the packing shed are, from left to right, Fred Arnold Bodily (obscured), Edward J. Tree, William Ogden, and Arnold D' Miller on the load. The rest are unidentified.

The Tree family, starting with Frederick John Tree in 1886, owned and operated this barn and 40-acre farm. Many Syracuse pioneers with the means and manpower built their own barns to assist with farming operations. Note the large hay derrick in the background; the farm primarily grew alfalfa or wheat.

Threshing, the act of separating edible parts of grain from the straw, was a tedious job. Many farmers used to cut the grain with a scythe and tie it into bundles by hand. The invention of horse- and steam-powered threshing machines allowed more work to be completed in a shorter time, yet it took a large crew to make sure the machine ran smoothly. Pioneer Orson Bybee purchased Syracuse's first threshing machine, seen here.

Farmers used hay derricks to move hay from a wagon or truck to a haystack. The derrick had a large log foundation with a sturdy upright in the middle about 10 to 12 feet high. On this upright, a long pole was attached on a swivel so that it could swing. A cable was attached to this pole, and on one end of the cable a Jackson fork was hooked to a hay mower pulled back and forth with a team of horses. Often, this was a young boy's job. (Used with permission, Utah State Historical Society.)

This photograph, taken around 1950, shows farmers harvesting and hauling hay from the field. The workers used an old hay hauler to load hay into the wagon. Alfalfa is first cut and dried for five days before harvesting begins. Alfalfa continues to be one of the most common crops grown in Utah.

The Thurgood family built this iconic barn in 1887. It stands to this day at 1000 West and 2300 South. Thomas J. Thurgood originally owned the barn and lived in it until 1925, when he sold the property to his son Joseph. Many settlers constructed similar barns, although most did not survive long into the 20th century.

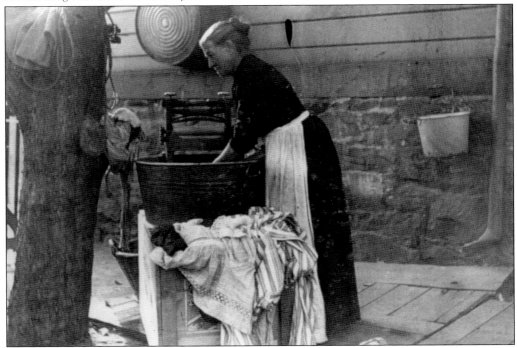

Wives took care of laundry and household chores while husbands worked the fields. Monday was always wash day. Multiple buckets and a washboard were used, and clothes always hung outside to dry. Roller presses helped speed up the drying process, but laundry still took a lot of time for pioneers. (Used with permission, Utah State Historical Society.)

Onions first settled their roots in Utah in 1914, when Aaron Call averaged between 1,580 to 1,920 bushels per acre on his farm. Unlike the sugar beet industry, the onion industry survived and even thrived after World War II. For many years, farmers harvested onions by hand, but over time, mechanized tractors took over. Onions require four months to grow from seed to crop. Above, trucks from Syracuse business Utah Onions Inc. haul loads of onions from the late summer harvest, undoubtedly to be shipped all over the United States. At right, an onion crop is ready for harvesting. (Both, Onions 52.)

The "Snow Horse," as the pioneers called it, remains a local landmark on the Wasatch Mountains. As the warmer seasons take over, melting snow creates an image of a horse for a short time between May and June. When the horse appeared around the 10th of May, pioneers knew the danger of frost was over. Based on the date of its arrival and length of its stay, the Snow Horse helped locals estimate how much water would be available for irrigation and whether it was time to plant tomatoes and peppers. Years when the Snow Horse did not appear resulted in summer droughts. (Diane Palmer.)

Three

FROM SMALL VILLAGE TO CITY

As Syracuse bustled with life, more people settled in the area. With irrigation and improved row crops, the community grew quite rapidly. With the addition of the Davis-Weber Canal, there was more irrigated land, with less acreage able to support more people. The next generation of Syracuse residents typically stayed in the area to continue the work that their families started. Around this time, Japanese families arrived to find work and start a new life in America.

Despite the Roaring Twenties raging through the rest of the United States, Syracuse was slow to adapt to the times. With population growth came problems. Water was easily contaminated, unsanitary sewage facilities plagued the area, and services were needed, such as public safety, fire protection, and others. In 1935, a group of townspeople came together to tackle these issues; in September of that year, they organized the Syracuse Town Board. This helped to facilitate further growth of Syracuse.

Town designation allowed businesses, service clubs, and other groups to form with a sense of camaraderie and pride. Some of these were the Lions Club, Lady Lions, sports leagues, and clubs for personal enrichment. Many still exist today and continue to enhance Syracuse.

United States of America

STATE OF UTAH, } ss.
County of Davis

I, Glen Day , County Clerk in and for
the County of Davis, State of Utah, and Ex-Officio Clerk of
the District Court, Second Judicial District, do hereby certify
the foregoing to be a full, true and correct copy of the

PETITION

for the incorporation of the

TOWN OF SYRACUSE

approved by the Board of County
Commissioners Sept. 3rd, 1935,

that I have compared the same with the original now remain-
ing on file in this office and that it is a correct transcript
therefrom and of the whole thereof.

IN WITNESS WHEREOF, I have hereunto set my hand
and official Seal this 3rd day of September A. D. 19 35

Glen Day
County Clerk.

By Ida Harty Deputy

As Syracuse grew, citizens realized the need for a city designation, which would aid in funding crucial amenities that an incorporated town would need. A petition was started in September 1935, and soon after, Syracuse was recognized as a city. The Syracuse Town Board was organized in 1935, with the first members, Wallace Christensen, Thomas J. Thurgood (president), Lionel E. Williams, Elton Bennett, and T. Joseph Steed, elected on September 11, 1935.

Born to pioneer parents, Thomas Thurgood was instrumental in the incorporation of Syracuse in 1935 and was the first mayor. He led the first town board, which helped provide benefits to Syracuse from state funds for new towns. He was president of the Central Mercantile Co. of Syracuse and director of the Kaysville Canning Corp. Here, he stands in front of his house that he built himself.

This home, which belonged to Joseph and Ruby Holbrook, hosted town board meetings from 1948 until Joseph's death in 1950. He served as the clerk for the new board, which had A.O. Stoker as president. According to board minutes, "his home was our town hall, where we have received every courtesy and welcome that big hearts could bestow." It was here that the change of Syracuse's status from town to third class city was discussed.

Syracuse's first post office was commissioned on November 16, 1891. Prior to that, mail was delivered through Clearfield. Clara Scholfield became Syracuse's first postmistress in 1903 when her husband purchased the Coles home and farm where the post office had resided. She operated the post office out of a small area with a little desk and window in the front of the house. Mail was delivered throughout Syracuse, Clearfield, Hooper, and Clinton. Service continued until May 15, 1905. With Rural Free Delivery, mail to Syracuse came through the Layton Post Office.

William Robert Allen was a well-known architect in Davis County. Many communities from Bountiful to Syracuse relied on his skills for projects ranging from bungalows to mansions and commercial buildings to religious centers. Allen even built his own home, which he lived in until his death in 1928. He designed and built three buildings in Syracuse: the Leo T. Warren home, the Criddle home, and the Syracuse Central School, which later became North Davis High School.

Leo Warren's family originally lived in the James Warren home, which was vacated after his father moved to Clearfield. In 1923, Leo decided to build his own home. Designed by architect William Allen, the bungalow at 2626 West 700 South had a sink and indoor bathroom with running hot water. Soon after the move, Leo became ill and was not expected to live. However, he pulled through and continued to farm the land near his home.

The Joseph Steed home is a typical Davis County homestead. Steed was a hardworking, well-known member of the early Syracuse community. He was an educator and farmer for nearly all his life. His home sat on 200 acres of farmland. During World War II, Italian prisoners of war were brought to the Steed farm to work. The farm was known for its roaming peacocks, and the home still stands to this day.

The Syra-Lita club was organized in 1937 during the late Depression years. Twenty-nine women formed this service club to help Syracuse transition to a town. The name was an abbreviation of Syracuse Literary and Service Club. Dues were $1 per year, and the club raised money for community projects, also donating time and items to local organizations. They had a book club on the side, and disbanded in 1993.

Recognizing the need for a cemetery in Syracuse, a cemetery committee was formed in 1896. Members selected a site on the farm of Alma and Catherine Tolman Stoker, who donated additional land in 1920. Later, more land was purchased from the Stoker family, bringing the total to almost seven acres. The first burials in the Syracuse City Cemetery were those of the young children of James Wood and Thomas Briggs: Lovemia (left), on July 17, 1896, and Delbert (below), on January 2, 1899.

The Syracuse cemetery was dedicated on Memorial Day 1940. Arnold D. Miller, president of the North Davis LDS stake, delivered a dedication address during the services. A history of the cemetery was given, including the story about the original need for a cemetery. Shortly after this ceremony, the cemetery expanded and become more meaningful to the community.

Today, the Syracuse City Memorial Park is the final resting place for many Syracuse citizens. The Parks and Recreation Department manages the site and grounds. It is tended to in all weather, and employees set up maps to help locate graves.

Horse racing became popular as Syracuse grew. Participants attached crop weights to their wagons to increase the challenge. Races were held in the warmer months when work was finished in the fields. Pictured here is Elton Briggs, a frequent race participant.

Eli Rentmeister started the famous Syracuse Boy Scout Band. Rentmeister, who moved to the United States as a young boy from Belgium, had a love for music. He led the band not only to teach music, but also to raise morale in the area. He brought the band to the beet fields so they could play for the farmers working all day. They also marched in several Fourth of July and Pioneer Day parades. This photograph was taken in 1921.

The Lions Club constructed a clubhouse to act as a local community building in 1970–1971. Wedding receptions, birthday parties, family gatherings, queen contests, and annual Christmas parties were held here. Helen Willey Barber, R.C. Willey's wife, provided the land and much of the interior decoration. Club members and others donated time and equipment to complete construction.

Receiving its charter from Lions International in 1951, with James Rentmeister as the first president, the local Lions Club club found many opportunities to provide recreational facilities to the small community whose government had barely been established. These included playgrounds, tennis courts, restrooms, bleachers, and new parks. Other significant activities included funding the publication of *History of Syracuse* by Cora Bodily Bybee, Fourth of July celebrations, and funding the barn building at the Syracuse Regional Museum.

The Syracuse Lady Lions Club was organized in 1963 with 21 members. The purpose of the organization, which still exists today, is to take an active interest in the civic welfare of the community and to aid where needed. The Lady Lions sponsored the Miss Syracuse Pageant, offered multiple service projects each year, and continue to raise funds for organizations in need.

The Miss Syracuse Pageant originated in 1955 when the Lions Club began a scholarship competition for young women to earn money for college. In 1963, the Lady Lions took over sponsoring the pageant. It ran until 1974. In 2000, it was brought back. Miss Teen Syracuse was added in 2005 in conjunction with Miss Utah/Miss America. Pictured is Miss Syracuse 1968 Verlynn Dahl (center), with first attendant Judy Thayne (left) and second attendant Marlene Garrett.

Baseball was strongly supported by the locals. All the shops and businesses closed on Saturdays in the summer because everyone was out watching or playing ball. When trucks and automobiles became prominent in Syracuse, they lined each foul line, and when a run was scored or a home run was hit, the fans clapped and horns honked for 10 minutes or more. Syracuse teams competed in statewide competitions and won many titles. Pictured is the 1916–1917 Farm Bureau team.

The Japanese people of Syracuse, who arrived in 1913, worked hard to improve the community through farming and business ventures. Like the rest of the community, they were interested in sports, mainly baseball. They would work hard during the week and support their team on Sunday afternoons. With assistance from Edwin Gailey, this team won its first championship in 1927.

Many Japanese migrated to Syracuse due to the Russo-Japanese War in the early 1900s. They became a huge part of the community, and loved playing baseball. They were the only Japanese population in Syracuse to not be sent to internment camps during World War II.

After the incorporation of Syracuse as a town in 1935, this sign was posted on Antelope Drive westbound toward Antelope Island. At the time of incorporation, Syracuse had a population of 835. This sign is currently on display at the Syracuse Regional Museum. (Author's collection.)

Four

IMPACT OF WORLD WAR II

Farming had always been the main means of employment for those living in Syracuse from the early settler days through the Great Depression. While some could work seasonally in the canning factories, beet dumps, or the sugar factory in Layton, most stayed local. Some of Franklin D. Roosevelt's New Deal projects came to Syracuse in the late 1930s as smaller projects such as forming cement foundations for outhouses, yet the impending war quickly clouded over Syracuse.

Many changes came as World War II broke out overseas. Hill Air Force Base (then called Hill Field) opened in 1940 and was a vital maintenance and supply base with round-the-clock operations to support the war effort. Wartime employment at Hill peaked in 1943, with over 22,000 military and civilian personnel. These dedicated men and women rehabilitated and returned thousands of aircraft to combat.

The enormous buildup in military supplies led to the establishment of a network of federal supply depots. Because of its central and somewhat isolated location, the government established many installations in Utah, including Army ammunition depots at Ogden (Ogden Arsenal) and a naval storehouse in Clearfield. Many Syracuse citizens worked at these locations.

With local men shipping off to war, Syracuse proudly contributed on the home front. This included planting victory gardens, saving every scrap of spare metal for ammunition and war materiel production, and doing everything they could to support their loved ones far away.

As World War II ended, the Cold War hit Syracuse beginning in the late 1940s to 1950s. The quiet, agrarian society transitioned to housing projects and industries that would soon dominate the economy. Long gone were the fields of sugar beets as farms slowly dwindled away.

The Ogden Air Materiel Area operated from 1934 to 1963. It provided program management and logistical support for the US Army Air Corps and later US Air Force. It also provided technical and logistical support for Air Force units in the western United States. Officials believed that inland depots were less vulnerable to air attack than coastal facilities. The Ogden facility provided many jobs to Syracuse citizens during World War II and the postwar era. These photographs were taken in 1955 (above) and 1957. (Both, used with permission, Utah State Historical Society.)

Established by the government in 1939, Hill Field was activated in 1940. It serviced and maintained aircraft to support the war effort. At its peak, upwards of 15,000 military and civilian employees enjoyed secure employment here, including many residents of Syracuse. The base has had a tremendous impact on the local and state economy. It continues today as Hill Air Force Base. (Both, used with permission, Utah State Historical Society.)

In the spring of 1941, the decision to build an inland naval supply depot in neighboring Clearfield had an immediate impact on Syracuse. Choice farmland was taken out of production, and huge warehouses were built. Many job opportunities opened, and people had cash in their pockets for the first time since the Depression. These aerial photographs, taken 20 years apart in 1943 (left) and 1963 show a portion of the huge storehouses, which stocked almost half a million items. During the postwar years, the Clearfield depot warehoused surplused and unused materiel. When the government decommissioned the depot in 1962, it was sold to the State of Utah and currently operates as the Freeport Center.

Promoting victory gardens in local papers, as seen here in the *Weekly Reflex* of Davis County, was a form of home front propaganda that inspired local farmers and civilians to contribute to the war effort. Articles described "Battle Plans for War Against Garden's Enemy Agents," along with specific how-to manuals for gardening in the local area, and cartoons meant to boost morale. This cartoon, showing the entire family working hard on the field (note the father is missing as he is presumably off to fight in the war), shows strong children, happy faces, and a thriving garden.

During World War II, a large percentage of the meat and produce grown in the United States had to be shipped overseas to soldiers. Consequently, civilians at home had to raise and preserve as much food as possible. To answer their needs, people were asked to contribute by growing victory gardens. Here, a local woman tends her victory garden in 1944.

During World War II, Utah had prisoner-of-war camps, with six base camps and six branch camps, for 7,000 prisoners. The German camp was on the southwest corner of the naval supply depot, with part of the camp in Syracuse. Because there was a shortage of labor, prisoners of war were called upon to do farmwork and other labor. Prisoners were given cots and food, and were allowed to partake in recreational activities. At war's end, the camp shut down and the prisoners were released and returned home.

The Variety Anns Club organized in the summer of 1944. It consisted of young women whose husbands were in the armed forces and away during the war. They reviewed books, played games, and crafted together. They also undertook civic projects to raise funds for different organizations, such as cancer drives, the American Red Cross, and American Field Service, and helped needy families at Christmastime. The club disbanded in June 1991.

56

Ned E. Rentmeister entered the US Army Air Force in August 1943, becoming a sergeant and serving as an aircraft gunner. He was one of only two Syracuse men killed in action in World War II. A service was held for him in the Syracuse LDS ward chapel on December 17, 1944. Rentmeister flew more than 20 missions over enemy territory prior to being killed when his plane crashed into a mountain in Italy on October 14, 1944. Below, in October 1945, Brig. Gen. Ray G. Harris posthumously awards the Air Medal to Sergeant Rentmeister in a presentation to Rentmeister's six-year-old son Larry and his widow, Martha Irene.

While World War II did not officially end until shortly after the atomic bombings in Japan, Germany's surrender brought much celebration to Utah and the entire world. The unconditional surrender of the Third Reich was signed in the early morning hours of Monday, May 7, 1945, in northeastern France. The majority of Syracuse men who were shipped off to war were sent to the European theater, and many made it home safely.

Five

LOCAL BUSINESSES

The local agrarian economy called for businesses that mostly aided farmers. This included blacksmith shops, coal yards, canning companies, mercantile stores, and garages. However, with the outbreak of World War II and the changes brought about by the Cold War and other aspects of late 20th century life, all kinds of businesses materialized. Building codes and zoning laws, which emerged in later decades, dictated where certain businesses could be built.

A famous business with Syracuse roots is R.C. Willey, the furniture and appliances store found all over the western United States. Young Rufus Call Willey sold appliances door-to-door in the 1930s. His innovative method of bringing products to the consumer made him successful. Despite the lack of an R.C. Willey store in Syracuse (the original store closed and was torn down in 2020), the roots of the company's success were in Syracuse.

Many current residents remember the businesses of their childhoods. Jim's Sport Center, owned and operated by Jim Rentmeister and his family, was the local hangout for farmers and townspeople for years. Many would gather there early in the morning during pheasant hunting season, and parents even hid their childrens' Christmas presents there in December. Other popular businesses included the Central Mercantile, Utah Onions, Hamblin's Foodtown, and Kano and Sons.

Now, Syracuse is a sprawl of chains, big-box stores, and some local businesses trying to survive in the tough economic climate of the 2020s.

Formerly the Syracuse Mercantile Company, the Central Mercantile Company operated in 1930 under a new board of directors. Besides selling hardware and groceries, the store had a lumberyard and mill in the back where windows, doors, and cabinets were made. David Thurgood (below) was the manager, director, and treasurer for 31 years until 1959. The store was eventually remodeled and expanded in the 1940s to keep up with growth during the war. Many things were rationed, and ration books were a must to get meat and other groceries. Some articles became scarce and were not seen again until after the war.

In 1918, Eugene Tolman began work as a blacksmith in Layton. He moved to Syracuse in 1926 and purchased a half acre of land to open his own blacksmith shop. His business steadily grew until the late 1930s, when farmers began using tractors instead of horses. Tolman then reinvented his business to include auto and tractor repairs. His sons took over operation of the auto repair business in 1946, and he eventually sold it in 1964.

In 1964, the Kano family left their life of sugar beet and tomato farming behind and purchased Eugene Tolman's blacksmith shop. They changed the name to Syracuse Repair. Ray Kano operated as a blacksmith, repairing farm machinery. His two sons, Ivin and Calvin, repaired automobiles. They took the business over from their father upon his retirement in 1975 and renamed it Kano and Sons. The business closed in 2020.

After World War II, demand for homes and commercial buildings was rising in Syracuse. Dale Smedley decided to open Smedley Plumbing and Heating in 1947, and by 1952, he had expanded the company by installing sewer and water lines in subdivisions. The business thrived and even ventured into construction. By late 1963, Smedley established the Smedley Development Company, an offshoot of his original business.

While pictured as Glenn's House of Meats, this building originally housed the local cheese factory in 1925. It was built by Roy Freed, who hired a Swiss cheese maker to operate the business for a few years. It was successful but closed in 1933. People could get more for their milk by taking it to Ogden and selling it for class A milk rather than class C, which was used for making cheese. In 1945, the Feller Packing Company of Bountiful purchased the cheese factory. Later, Glenn Buhler sold wholesale frozen meats and Whirlpool freezers from 1961 to the mid-1980s.

Rufus Call (R.C.) Willey was a prominent businessman in Syracuse. In 1932, he sold appliances door-to-door out of his truck to supplement his income. Eventually, his business grew as electric refrigerators and ranges became popular. Willey opened his own store in 1950 next to his house, hired a full-time employee, and established a telephone line from his home to his business. Things went well until he died of cancer in 1954. After this, his wife, Helen, asked her son-in-law William "Bill" H. Child to run the business. Bill brought his younger brother Sheldon into the company, and together they built a reputation for outstanding customer service and great values. By 1993, R.C. Willey had grown to five retail stores, and the corporate office moved from Syracuse to Salt Lake City. The photograph below shows the updated Syracuse R.C. Willey built in 1994. It eventually closed, and the building was demolished in 2020.

The last dairy farm in Davis County, Hamblin's Dairy Farm, originated in 1904 and closed in 2013. The farm produced milk, cheese, and even ice cream. Five generations of the Hamblin family operated it for over 100 years. Current Syracuse citizens fondly remember visiting the farm on school field trips.

George and Marion Hamblin bought the Thurgood's Market building on January 7, 1980, enlarged it, and turned it into Hamblin's Food Town. The drive-through was a popular addition to the store as people enjoyed the quick access for purchasing a gallon of milk or a loaf of bread. School children flocked to the store after classes to buy or trade in their bottle caps for penny candy. The Hamblins eventually sold the store in 1994, which was turned into Western Market. However, with Smith's Food and Drug opening in 1996, Western Market quickly went out of business.

Jim Rentmeister, a tireless advocate and figurehead for Syracuse, owned and operated Jim's Sport Center. Established in 1947, it provided a plethora of sporting goods, diner food, and a space for locals to hang out. It was a gathering spot for farmers to get a cup of coffee and catch up on the latest news and gossip, and the hub for autumn duck and pheasant hunting. Jim's twin daughters, Pam and Paula, and their older brother Alan also worked in the café. It was well known for its homemade hamburgers and delicious milkshakes. Pictured below are Pam and Paula lounging inside the café during a slow shift. At right is the Jim's exhibit at the Syracuse Regional Museum.

Roy Miya's garage was the first Japanese-owned business in Syracuse. Miya started his garage in 1949 and operated it for 36 years until his retirement in 1985. He and his brother Kazuo provided mechanic work and kept busy from Syracuse patrons and Job Corps fleets. The business was bought in June 1991 by Paul McBride, a native Syracuse citizen who continues to operate Paul's Automotive today.

Roy Miya was born on December 12, 1920, in Roy, Utah, the son of Gontaro and Soyo (Kawaguchi) Miyahishima. He graduated from Davis High School in 1939 and went to National Auto School in Los Angeles, California. He returned to Utah and not only farmed the area, but also owned and operated Miya Garage in Syracuse. Miya dedicated much of his time as a volunteer firefighter with the Syracuse Fire Department. On January 5, 1966, he was appointed by city council as the new fire chief, with Val Cook elected by the firefighters as his assistant. Miya served for 19 years, retiring on the last day of December 1984.

Terry Palmer and Michael McBride started Star Video in 1984 in this building. The store primarily distributed videos and VCRs to remote towns in Utah, Idaho, and Wyoming that were too small to have their own video stores. Star Video owned about 7,000 videos and was a weekly stop for many locals. Palmer and McBride owned the business for 11 years before selling to Rick Cowley, who had operated the store for them. (Diane Palmer.)

Stoker's Nursery was a family-run business that started in 1967 at 2050 South 1000 West. The business won the Overall Achievement Award in 1992 from the Fred Meyer Co. for excellent service and quality. It was one of three nurseries in Utah awarded for that level of service. Stoker's ran for many decades until it closed in 2012.

In October 1955, Max Waite passed his journeyman's barber test and opened Max's Barbershop in Syracuse. He cut hair for over 50 years, and visiting his shop became a common pasttime. In addition to being a barber, Waite was a farmer, volunteer fireman, bus driver, and justice of the peace for Syracuse. He was a well-loved member of the community.

Max's Barbershop was originally on the corner by the Central Mercantile Company (1700 South and 2000 West). In 1976, Waite moved the shop closer to his home at 1275 South and 2000 West, sharing the space with his daughter-in-law Dixie. The barbershop closed in the 1990s. In 2006, it was relocated to the Syracuse Regional Museum and has been preserved there since. It took careful planning by the city and the museum to move the building while maintaining its integrity. Stepping inside feels like a trip to 1960s Syracuse—even the barbershop smells are still there.

Utah Onions Inc. started business in 1977 and had 30 successful years growing and selling onions in Utah. While a tough crop to grow, onions thrive in Utah. Over time, Utah Onions had national sales, and it became the largest onion grower in the United States. In 2007, the company was sold and renamed Onions 52, which expanded operations to nearby states. Onions 52 is still in business, and the beloved Utah Onions sign is on display at the Syracuse Regional Museum.

Six

EDUCATION AND FAITH

Early settlers knew the importance of education. In 1885, a one-room schoolhouse was constructed on the southeast corner of 4000 West and 1700 South. All eight grades had class together. Textbooks were scarce, but the students learned reading, writing, and arithmetic. Unfortunately, most children missed many school days to help on their families' farms. A student who attended five months of school in a year was lucky.

A little brick schoolhouse was built on 700 South and 2000 West and became the first high school in Davis County in 1909, aptly called North Davis High School. It was an important element in the lives of the young people of Syracuse, Clearfield, Clinton, and West Point. The school sponsored many recreational activities outside of education, such as sports teams, a band, and a meeting place for programs for those wanting to continue their education through adulthood. As Syracuse grew, so did the need for schools, and now the bustling suburb has 20 schools and counting.

The Church of Jesus Christ of Latter-day Saints was the dominant faith of those who settled here in the mid-1800s. In 1877, the Davis County Stake was organized with William R. Smith as president. In 1895, the Syracuse Ward was organized with David Cook (who plowed the first furrow in Syracuse) as bishop. In 1897, a brick chapel was built at a cost of $12,000. Unfortunately, it was destroyed by a fire on October 3, 1940. This was the only ward in Syracuse until 1953, when the Second Ward was created with Mark Beazer as bishop.

Today, the church is building its first temple in Syracuse, which will be its third in Davis County and twenty-fourth in Utah. It is scheduled to be completed in 2024.

These photographs show students at the Syracuse School in the 1890s. Most walked or rode their family horse to school. Horses were kept in a stable beside the building. These students occupied a one-room schoolhouse until it was expanded in early 1913. They studied reading, writing, and arithmetic, although work on the farm sometimes kept them from attending.

This was the first official school bus in Syracuse. Children who lived more than a mile from the school had the option to be picked up and taken to school. The bus would drive around Syracuse and West Point to pick up kids and haul them to North Davis High School. This 1926 photograph is the only one of this vehicle.

By the early 20th century, the Syracuse School District had grown to 377 children. The need for more schools was urgent. However, students were able to get a better education once schools separated into elementary, junior, and high school. The attire seen here in 1911 was common for graduating seniors.

North Davis High School was the first high school in Davis County. Housed in the original small redbrick schoolhouse built by architect William Allen in 1909, it was enlarged and opened in the fall of 1911. By the end of 1914, the school was growing with an active, enthusiastic student body, and another room was added. Each year, the school had a prom, play, operetta, and a domestic science fair. They competed with the larger schools and even beat Davis High School in basketball one year. Unfortunately, North Davis High closed after the 1924–1925 school year due to district consolidations.

A play called *The Spinning Wheel* was presented at North Davis High School in 1920–1921 as part of a local competition against other high schools. North Davis High was not only the first high school in Syracuse, but also in Davis County.

As schools expanded along with the population in Syracuse, so did extracurriculars. One popular activity was band, as a love for music was passed down by pioneer parents and grandparents who would played instruments after a hard day on the farm. Band instructor Lester Gleason is pictured here with, from left to right, (first row) Golden Waite, Ben Thurgood, Max Waite, Blair Barber, Verl Barber, and John Knighton; (second row) Lola Hammond, Beverly Bennett, Donna Bodily, George Fisher, Ferrel Gailey, Lee Schofield, Gary Cook, and Milton Wilcox.

Syracuse Elementary School was on the east side of 2000 West just north of 1700 South. Although several schools existed in the area by the early 1900s, there was a need for more due to growth in the city. The school expanded multiple times throughout the 20th century to add classrooms and other amenities, such as a library, restrooms, and a central furnace room. The original building and historic facade were eventually torn down in August 1982. While there is still an elementary school at that location, it is a new, updated building.

Japanese students are presenting a cultural dance in front of North Davis Junior High School during an event in 1940. From left to right are Isao Yokomizo, Shizue Shimoyi, Haniyi Endo, and Yuki Yaneda.

Banschichi Kano was one of the first Japanese Americans to settle in Syracuse in the early 20th century. He was born in 1877 in Japan and immigrated to the United States in 1900, marrying Ito Nomura in 1913. They had five children: Ray, Russel, Kayzo, George, and Larene; many of them grew up to serve their community and start their own businesses. When Banschichi Kano passed away in 1944, most of Syracuse attended his funeral.

Many Japanese settlers practiced the Buddhist faith. Plans to build a Buddhist church in Syracuse began in 1923 through the efforts of the Salt Lake City Buddhist Church. Land was acquired on the Barnes Farm, and a wooden structure about 20 feet by 40 feet was constructed. It was completed and dedicated on April 26, 1925. This building was used for Japanese language school, farm organization gatherings, social groups, movies, judo, and church services until March 1980, when the Syracuse Buddhist Church merged with the Ogden Buddhist Church. The building was demolished in 1990.

This brick chapel was constructed in 1897 for about $12,000. On October 23, 1940, the Syracuse Ward Chapel caught on fire and burned to the ground. Ward functions were held in the amusement hall until a new chapel was built. On May 24, 1942, a new chapel was completed at a cost of $54,000 and dedicated by president David O. McKay, who was second counselor in the first presidency. Over 1,000 people attended the dedication. (Used with permission, Utah State Historical Society.)

The Syracuse Ward Chapel was built in 1942 after the original brick building burned down. Some of the bricks from the original chapel were salvaged and used in the new building. It was the only ward and stake center in Syracuse until a second stake was created in Syracuse in 1952.

Seven

THE GREAT SALT LAKE

The Great Salt Lake remains one of the most intriguing natural phenomena in the world. The Native Americans who inhabited the area trusted its healing powers, and it did not take long for the pioneers to discover its buoyant waters where a person can float like a cork. Pioneers also sailed upon the dark blue waters to discover numerous desert islands. Many historic accounts recall the glorious, awe-inspiring sunsets of pink, yellow, purple, and orange hues on the lake.

Salt harvesting became a profitable business venture along the shores of the lake. While many of these enterprises popped up along the entire shoreline, William Galbraith's business is credited with naming the town of Syracuse. Adopted from the salt mining industry of Syracuse, New York, Galbraith called his business Syracuse Salt Works. The name stuck.

The Syracuse Bathing Resort opened in 1887, which brought in crowds from Ogden to Salt Lake City. It was an oasis in the desert, with people enjoying beautiful days on the lake, dancing at the state's largest dance pavilion, and relaxing after long days of working on the farm. Arguably, the resort brought many newcomers to the area. Unfortunately, the oasis was short lived, as it had to shut its doors before the turn of the 20th century.

Antelope Island, the largest island in the Great Salt Lake, became a homesteading property run by the Frary family in the 1840s. Many other families came to live on the island, which was only reachable by boat. In 1969, the state bought the island and established it as a state park. Eventually, a causeway was constructed in 1969, connecting Syracuse to Antelope Island.

The Great Salt Lake, as with any large body of water, can be unpredictable. In 1983, the causeway was severely damaged due to flooding, and it took until 1992 to completely fix. Water levels vary between historically high (1987) to historically low (2022) depending on precipitation levels. In March 2023, thanks to historic record-breaking snowfall, the lake increased by three feet, giving locals hope that drought conditions might not ruin the precious Great Salt Lake.

Harvesting salt off the Great Salt Lake was once very common. The first person to start a salt business in Syracuse was George Henry Payne, who later sold his salt ponds to William Galbraith.

While the Great Salt Lake does not typically freeze due to its high concentrations of salt, sometimes during calm winter weather, fresh water from flowing streams can mix with lake water and freeze some sections. This has sometimes caused an ice sheet several inches thick to extend from the Weber River west to Fremont Island. In the early 1900s, one such ice sheet made it possible for coyotes to cross to Fremont Island and attack sheep pastured there. The breakup of thick ice has also been known to form icebergs. One iceberg in 1942 was 30 feet high and 100 feet wide. (Used with permission, Utah State Historical Society.)

Great Salt Lake ice floes form as fresh water freezes at mouth of streams feeding lake, are blown into huge mobile piles, some 30 feet high.

Great Salt Lake 'Iceberg'

Wrecked a Truck

On the 'Highway'

To Fremont Island

by David E. Miller
History Department, University of Utah

Floating ice menace roughed up Charles Stoddard and his "Lakemobile."

"THIS IS ABOUT the spot where I was struck by an iceberg back in 1942! I was driving a truck to Fremont Island. . . . Yes, it was right out here in the middle of Great Salt Lake." That's what Charles Stoddard of West Point, Utah, told me as we were cruising the lake some time ago.

Now, everyone knows the Great Salt Lake is too salty to freeze, even in the coldest winter weather. Anyway, what was Stoddard doing in a truck out in the middle of the lake?

It surely sounds like a ridiculous statement— for a truck to be struck by an iceberg in the middle of a lake that never freezes. Yet what he told me that day is absolutely true.

This is how it happened: During cold, calm winter weather fresh water at the mouth of the streams that feed the lake literally floats on top of the dense salty brine and sometimes freezes into ice several inches thick. This ice often drifts about the lake in huge sheets, at times doing damage to boats and shore installations.

Editor's Note: This is one of a series of articles on pioneers, places and events prepared under the direction of the National Society, Sons of Utah Pioneers, in the group's "Know Your Utah" campaign.

In heavy wind storms these sheets of ice are broken up and sometimes blown into huge piles that have all the appearance of icebergs. But it is the floating sheets of ice that cause most of the damage.

Of course, these are not true "icebergs" and should more accurately be called ice floes. If a truck happens to be caught in the path of such an ice floe, that vehicle stands a good chance of being destroyed. But what would a truck be doing in the middle of the lake?

For several decades Charles Stoddard has been using Fremont Island as a sheep range. One of the difficult problems of the business has always been that of transportation—for sheep, wool and men.

When the lake was extremely low during the late 1930s, Stoddard solved this problem by locating a submerged sandbar that extended southward from Fremont Island and angled off toward Syracuse Point. By following this bar he found that he could easily ride horseback to and from the island. Wagons could be taken by the same route.

THE NEXT STEP was to plant posts along the route to keep a lake navigator from going astray—getting off the sandbar and into the soft lake mud. (One driver tried a shortcut and lost a fine team of horses in the quicksand lake bottom.)

Since horse and wagon travel was too slow for the sheep business, Stoddard invented a lake-going truck—a "Lakemobile." This was accomplished by constructing some large caterpillar-like cleats as special equipment for the rear wheels of an "A" model Ford truck.

To increase the capacity of the cooling system, a large Fordson tractor radiator was added. When completed, this novel machine proved entirely lake-worthy and Stoddard could traverse the 10 miles of salt water trail between Syracuse Point and Fremont Island in short order.

While thus traveling from the island early in the spring of 1942 this machine was struck broadside by a slowly moving sheet of ice several inches thick and a half-mile wide. Stoddard felt certain that his Lakemobile was lost for sure this time.

The heavy ice plowed into the vehicle, tipped it up on two wheels and almost over before the truck slipped enough to allow all four wheels to settle back into the water. The driver then jumped out and, with the aid of a crowbar, managed to turn the machine around so that it could roll in front of the ice. He let it drift away, escaped on the ice, and returned home.

TWO OR THREE months later Stoddard rode horseback into the lake in search of his Lakemobile. He found it about a mile and a half south of the spot where it had been struck by the ice and abandoned. After changing the salt-impregnated oil, removing the spark plugs and pouring in a small amount of kerosene to loosen the cylinders, Stoddard had no trouble in starting the engine and the Lakemobile pulled out under its own power to be used a few more seasons in lake navigation.

During the past several years, with the gradual rise of lake waters, Stoddard has had to find other means of conducting his ranching business on Fremont Island. But for a few years he really had some rare experiences as he cruised the lake in a home-made Lakemobile.

This article from 1956 describes the strange and rare occurrence of icebergs in the Great Salt Lake. While the lake cannot completely freeze, during cold, calm weather, fresh water feeding into the lake floats on top of the denser salt water and sometimes freezes into ice several inches thick. These are not true icebergs but rather ice floes. The lake's salinity alters with changing weather conditions. On average, the Great Salt Lake is usually six times saltier than the oceans. While the lake is mostly inhospitable to boaters, swimmers have long been fascinated by its ability to easily keep them afloat.

The Syracuse Bathing Resort was described as an oasis in the desert not only by the people of Syracuse, but by many others from surrounding communities. It had 4,000 to 5,000 fruit trees, a beach with bathhouses, picnic tables in shady spots, and was landscaped with greenery and flowers. (Special Collections Department, Stewart Library, Weber State University.)

The Syracuse Bathing Resort opened on July 4, 1887. It was a popular spot for locals and people all over the valley. The spur rail line to Syracuse brought thousands of people a day from Ogden and Salt Lake City to the Great Salt Lake. (Special Collections Department, Stewart Library, Weber State University.)

Visitors hang out on the shores of the Great Salt Lake. This was the location of the Syracuse Bathing Resort, where many locals and visitors came to relax. Dress was modest—both women and men were fully clothed when swimming in the lake. (Used with permission, Utah State Historical Society.)

With the opening of the Syracuse Bathing Resort on July 4, 1887, thousands of people flocked to Syracuse to unwind. Locals loved living by the lake, and its beauty enhanced the community. Most guests enjoyed swimming and floating in the Great Salt Lake, especially on hot summer days. While it unfortunately closed in 1891, the resort's effects were long lasting upon the community as it attracted many permanent settlers to the area.

There have been many attempts to cross the Great Salt Lake by different means—boats of all sizes, swimming, or even walking if the water level was low enough, which has sometimes happened. People have long used boats to cross to Antelope Island, even using rafts to transport livestock and supplies. Some attempts to cross by vehicle resulted in minor catastrophes as they got stuck in the salty mud. (Used with permission, Utah State Historical Society.)

Bison were introduced to Antelope Island in February 1893 when White and Sons Co., which managed the island for the LDS Church, purchased 17 bison from William Glassman of Ogden (who had purchased them on a whim during a trip to Texas). Bison hunting was frequent until public sentiment changed during the 1920s and activists began to call for protection of the herd on the island.

This postcard from the 1930s shows a small herd of bison grazing along the grassy slopes of Antelope Island. Currently, the herd is one of the largest and oldest publicly owned bison herds in the nation, numbering between 550 and 700 individuals. It is one of two bison herds managed by the State of Utah, the other being the Henry Mountains herd. (Used with permission, Utah State Historical Society.)

Alice Frary, the wife of George Isaac Frary, settled her family on Antelope Island and took up homesteads in 1891. In September 1897, she became desperately ill. George sailed across the lake to get needed medicine in Ogden. When he returned to Syracuse he saw three fires on the island, which meant distress. Sailing quickly back to the island, his small raft capsized, and the medicine was lost in the lake. He made it back in time to tell Alice farewell, and she died on September 3, 1897. She was buried near their house at the end of the orchard with a small pink rock for a marker.

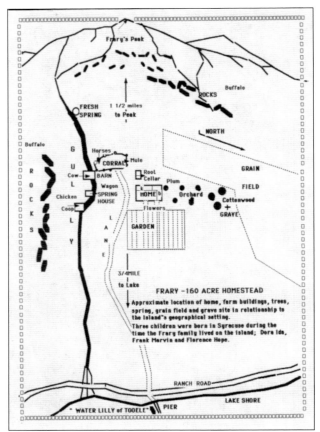

Within the map:

Frary's Peak

Buffalo
ROCKS

FRESH
SPRING

1 1/2 miles
to Peak

Buffalo

NORTH

G
R
O
C
K
S

U
L
L
Y

Horses
CORRAL
Cow BARN
Chicken
Coop
SPRING
HOUSE
Wagon
Mule
Root
Cellar
Plum
HOME b
Orchard
Flowers

V
A
L
L
E
Y

GARDEN

GRAIN

FIELD

Cottonwood
+
GRAVE

3/4 MILE
to Lake

FRARY -160 ACRE HOMESTEAD

Approximate location of home, farm buildings, trees,
spring, grain field and grave site in relationship to
the island's geographical setting.

Three children were born in Syracuse during the
time the Frary family lived on the island; Dora Ida,
Frank Marvin and Florence Hope.

RANCH ROAD

LAKE SHORE

"WATER LILLY of TOOELE"   PIER

This map outlines the homestead of the Frary family, who occupied Antelope Island for many years. George Frary moved here with his family in 1891 to take up a homestead. They built a three-room cottage, and Alice Frary educated the children while George farmed and sailed the Great Salt Lake. Eventually, the children left the island to pursue a life in Salt Lake City, while George continued on the lake with his beloved sloop, the *Water Lily of Tooele*.

Pictured here is the north face of Frary Peak, viewed from the visitors center. This peak was named after the last family to homestead the island, who settled just below the tallest peak. George Frary died at age 88 in 1942 after spending half a century on Antelope Island, longer than anyone else. Today, many hike up 6,596-foot Frary Peak to enjoy the majestic views of Salt Lake City, the Wasatch Mountains, and the Great Salt Lake. (Author's collection.)

The Syracuse Historical Commission coordinated the building of a monument at the Garr Ranch House honoring the families who lived and ranched on Antelope Island. It was dedicated on June 5, 1993. More than 500 descendants of the Garr, Stringham, and Walker families crossed the causeway from Syracuse and attended the event. On July 1, 1993, the new bridge was finally completed, and the Syracuse causeway was opened to the public.

Many people have enjoyed the Great Salt Lake. Swimming and sunbathing were popular with early Syracuse citizens, especially during the hot summer days without air conditioning. The salinity of the water averages about 12 percent, making it much saltier than the ocean. Pictured is Diane Palmer's family enjoying a soak in 1997. (Diane Palmer.)

Just as the state purchased Antelope Island in 1981 to develop it into a state park, Mother Nature had other plans. The year 1983 provided record-breaking levels of precipitation, causing floods and tremendous damage. With no outlet, the Great Salt Lake continued to rise. Needless to say, the Syracuse causeway was swallowed up. It was not until the early 1990s when locals and tourists could enjoy the island again. (Diane Palmer.)

After being closed for nine years, Antelope Island finally reopened. The ribbon was cut at the Syracuse entrance on Saturday, October 31, 1992, by state and county officials. This photograph, taken in May 2023, highlights the birdhouse (right) and causeway (left) curving back to Syracuse. (Author's collection.)

Eight

ENTERING A NEW CENTURY AND BEYOND

The 1980s–1990s in Syracuse have been called its "Golden Years" by longtime residents. Life was simpler—before the internet, smart phones, September 11, 2001, and severe political divisions that affect American life today. Many recall the vast open farmlands, buying penny candy and sodas from the local store, watching tractors haul sugar beets away, playing and fishing in the ditch water, riding bikes or horses everywhere, and a small, tight-knit community where everyone knew each other.

World events impacted Syracuse in the standard way. When terrorists crashed airliners into the World Trade Center and the Pentagon on September 11, Syracuse citizens were troubled, but stood united. Many youth time enlisted in the military and went off to fight in the War on Terror.

The COVID-19 pandemic hit Syracuse hard in March 2020. Like most of the world, Syracuse shut down while people quarantined in their homes for over three weeks. Shelves at local grocery stores were emptied, and people migration toward remote learning and remote work. While restrictions gradually eased over time, the ordeal lasted a little over two years.

In 2000, the census recorded Syracuse's population at 9,398. In 2010, it had increased to 24,331. By 2020, it reached 32,141. Thousands more have moved to Syracuse in the three years since. The increase in citizens called for an increase in infrastructure. Once a town with no stoplights, Syracuse is now a bustling urban area with rising apartment complexes and restaurant chains. Gone are the days of unsupervised kids wandering through town and a more relaxed way of life.

On September 13, 1950, a proclamation was signed by Gov. J. Bracken Lee that entitled Syracuse to become a third class city. This demonstrated a need for an actual city hall, and the first one was constructed in 1953. It was a small, unassuming building. The police department was around the back on the right, and the fire department was on the left. It served the community for many years until a larger space was needed. (Used with permission by Diane Palmer.)

As Syracuse grew over the last decade, so did the need for a larger city hall building with more space for offices. The new city hall was built in 2008 and has a full-time post office, council chambers, and provisions for more space when further expansions become necessary. (Author's collection.)

Darren Parry is the former chairman of the Northwestern Band of the Shoshone Nation and was born and raised in Syracuse. His grandmother Mae Timbimboo Parry was a storyteller and activist for their nation and brought awareness to one of the largest massacres of Native Americans in history, the Bear River Massacre, which took place just over the border in Franklin County, Idaho, in 1863. Darren is continuing her work to bring awareness by building the Wuda Ogwa Cultural Interpretive Center. (Darren Parry.)

DeLore Thurgood was an active member of the Syracuse community for many years. He was mayor of Syracuse from 1985 to 1991, board member and treasurer of the Air Force Heritage Foundation, and a volunteer for Friends of Antelope Island, then a member of the board of directors. In 1994, he was asked to head up the Syracuse Historical Commission Board. Once the Syracuse Regional Museum was established in 2002, Thurgood was the first president of the Syracuse Museum Foundation. He was responsible for new and exciting projects to improve the community.

This Syracuse Historical Commission was established in 1988 after realizing the importance of preserving Syracuse's heritage at the Constitution Bicentennial Celebration in 1987. This commission was responsible not only for jump-starting efforts for the Syracuse Regional Museum, but also wrote and published *The Community of Syracuse* history book. From left to right are (seated) Elaine Nance, Etheleen Holt, Clayton Holt, Donell Hansen, Louise Simpson, and Marian Hamblin; (standing) Don Rentmeister, Genene Rentmeister, Allen Willie, Helga Willie, Vaughn Hansen, Joe Simpson, and George Hamblin

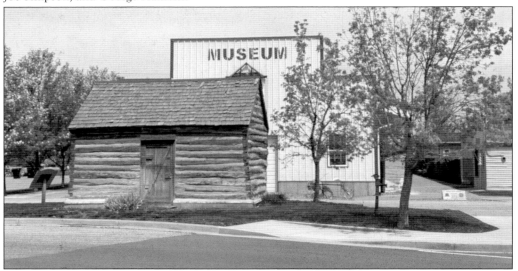

Since its opening in 2002, the Syracuse Regional Museum has grown and flourished. In addition to a main building, there is also a barn building, the Max Waite Barbershop, and the Wilcox Cabin. In November 2022, a mural was painted on the west side of the main building, making it the first mural and public art piece in Syracuse. It continues to present the history, settlement, and culture along the Great Salt Lake through exhibits, texts, and community outreach programming. (Author's collection.)

Heritage Days every June celebrates the history of Syracuse. Organized by the city, there are multiple events throughout the week as well as a Saturday morning parade. Heritage Days continues to be a popular event for the community. The Syracuse Regional Museum participates every year with a special artifact, while volunteers dress in period clothing. Pictured here is former museum curator Elizabeth A. Najim with a restored buggy that was in the parade in June 2021.

The Syracuse Utah Temple is currently under construction. It will be the first LDS temple in Syracuse and the third in Davis County. Estimated completion is mid- to late 2024. The building will take up 88,886 square feet and sits on 12.27 acres. There will be a single spire 319 feet tall, making it visible from Interstate 15. The temple will have two baptistries, four instruction rooms, and four sealing rooms. (Author's collection.)

This relatively new town square was constructed at the northwest corner of the intersection of Antelope Drive and 2000 West in 2005. The strip mall shells and the clock tower went in around that time, and businesses have slowly moved in over the last few years. The area continues to grow as new restaurants and stores move in. (Author's collection.)

COVID-19 hit Syracuse in March 2020, much like it did on an international scale. On March 11, 2020, Pres. Donald Trump announced the prohibition of incoming flights from Europe to the United States. Grocery store shelves were bare as people rushed to buy essentials such as bread and toilet paper. This was the bread aisle in Smith's Food and Drug in Syracuse at the time. (Nike Peterson.)

BIBLIOGRAPHY

Arave, Lynn, and Ray Boren. *Great Salt Lake*. Charleston, SC: Arcadia Publishing, 2022.

Bybee, Cora Bodily. *History of Syracuse*. Springville, UT: Art City Publishing Company, 1965.

Holt, Clayton. History of *Antelope Island: 1840 to 1995*. Syracuse, UT: The Syracuse Historical Commission, 1994.

Syracuse Historical Commission. *The Community of Syracuse: 1820 to 1995 Our Heritage Centennial Edition*. UT: Cedar Fort Inc., 1994.

Tucker, J. Kent. *An Examination of the Mormon Settlement of Syracuse, Utah*. Provo, UT: Master of Arts thesis, Brigham Young University, 1987.

DISCOVER THOUSANDS OF LOCAL HISTORY BOOKS FEATURING MILLIONS OF VINTAGE IMAGES

Arcadia Publishing, the leading local history publisher in the United States, is committed to making history accessible and meaningful through publishing books that celebrate and preserve the heritage of America's people and places.

Find more books like this at
www.arcadiapublishing.com

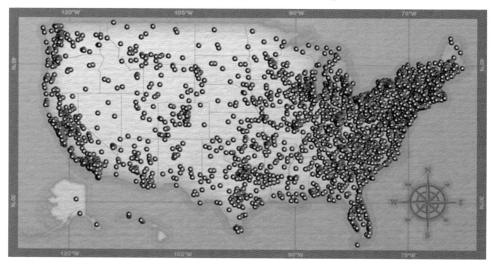

Search for your hometown history, your old stomping grounds, and even your favorite sports team.